Histology Hacks

Michael Backhus B.A HT (ASCP)

To: Kristin, Kylee, Kaden, Kassidy, Kaleb & Kamryn —

I will always be there for you, as you are me…..

thank you

Table of Contents

1.0 Introduction --- 14

2.0 Embedding --- 18

Air Bubbles --- 19

   2.1 Bubble Trouble -- 19

Hard Tissue -- 20

   2.2 Foreign Body/Hard Tissue in a Block ---------------------------------- 20

   2.3 Polar Ends --- 21

Lost Tissue -- 22

   2.4 Find & Go Scrape -- 22

   2.5 Jumper --- 23

   2.6 Meshy -- 24

  2.7 Orientation Station --- 26

Paraffin Prevention -- 27

  2.8 Baby Oil --- 27

  2.9 Edward Icy Hands --- 28

  2.10 Square Cold Spot --- 29

Stability of Blocks --- 31

  2.11 Refill -- 31

Tissue Size (Big to Small) --- 32

  2.12 Magnify -- 32

  2.13 Something Blue -- 33

  2.14 Tissue House --- 35

Tissue Stuck to the Sponge --- 37

  2.15 Taper Saver --- 37

Under Processed Tissue --------38

    2.16   Chamber Saver --------38

Uneven Planes --------39

    2.17   Re-embedding blocks --------39

    3.0   Microtomy --------42

Bubbles in a Paraffin Ribbon --------43

    3.1   Bubble Bath --------43

    3.2   Bubble Pop --------44

    3.3   Penny Drop --------45

Cysts --------46

    3.4   Cyst Concealer --------46

    3.5   Vicks Vapor Rub --------47

Decal --------48

3.6   Backflip Blocks -- 48

3.7   Lazy-Cal -- 49

3.8   Save the Decal --- 50

Dry Blocks --- 51

3.9   Moist Towelette --- 51

3.10 Moisturize -- 52

3.11 Warm Dip --- 53

Hard Tissue -- 54

3.12 Liquid Fabric Softener --- 54

3.13 Lotion -- 55

3.14 Pop Goes the Block --- 56

Identification of Block History -- 57

	3.15	Sharpie Solution -- 57

Microtomy Station -- 59

	3.16	Lockless Microtome -- 59

Nails --- 60

	3.17	Nair It -- 60

Optimal Water Bath Temperature -- 62

	3.18	On the Rocks --- 62
	3.19	Whistle While You Work --- 63

Ribboning Issues --- 64

	3.20	Finger Press -- 64
	3.21	Pencil Unsharpener --- 65

Soggy Tissue -- 66

	3.22	Cling Wrap --- 66

3.23	Foil	67

Static ---- 68

3.24	Aluminum	68
3.25	Damp Cloths	69
3.26	Dryer Sheets	70
3.27	Humidify	71

Tissue Placement on Slide ---- 72

3.28	Glue Trick	72
3.29	Quicker Picker Upper	73
3.30	Sliding on a Slide	74
3.31	Slip & Slide	75
3.32	Two Sections on One Slide	76

| 3.33 | Upside Down Slide | 77 |

Under Processed Tissue ---78

3.34	Frozen-less Section	78
3.35	No Explode Tissue	79
3.36	Suds in a Water Bath	80

Wrinkles ---81

3.37	Alcohol Bath	81
3.38	Ammonia	82
3.39	Double Dipping	83
3.40	Runway	84
3.41	Wrinkle Release	85
4.0	Frozen Sectioning	88

OCT Compound -- 89

    4.1   **Frozalin (Frozen to Formalin)**--- 89

Sectioning -- 90

    4.2   **Fog It Up**-- 90

Static -- 91

    4.3   **Cold Static**-- 91

    4.4   **Cry-O-Static**--- 92

    5.0   **Grossing**--- 94

Cleaning & Disinfecting Grossing Utensils-- 95

    5.1   **Dreft** --- 95

Epidermis Error --- 96

    5.2   **Ellipse Skin Fragment** -- 96

Filter -- 97

 5.3 Cap Full -- 97

 5.4 Forma-Dip -- 99

 5.5 Squeeze It -- 100

Prevention of Cross Contamination, Inaccuracy, and Disorder of Specimens ----- 101

 5.6 Baskets --- 101

 5.7 Colors of the Eight Pack --- 103

 5.8 Even & Odds -- 104

 5.9 Histo-Pic -- 105

 5.10 Triple Check -- 106

Safety --- 107

 5.11 Safety Blade -- 107

Tiny Specimens --- 108

5.12	Dip Stick	108
5.13	Orange is the New Black	109
5.14	Orange Options	110
6.0	Conclusion	114

Introduction

1.0 Introduction

I've documented and combined a great deal of tricks and tips that will help with your histology technician work. You can call these secrets or tricks of the trade. These are things you won't be taught in school and can't find in a textbook. They are on the job, years of experience, trial, and error. Some techniques I've created. Others I've learned at different laboratories in different states where I have visited or worked. I have communicated with other technicians and learned their techniques. I refer to these techniques as "hacks". Hacks are improvisational ways created to improve histology. I categorized these hacks into four divisions. I termed the four divisions embedding, microtomy, frozen sections, and grossing. I have personally demonstrated each "hack" and determined its effectiveness. I explain the science behind each technique. I go into depth on the four divisions of histology hacks. I think you will find this book intriguing and useful. It will help you produce better quality work, be more efficient, and hey, maybe even impress your coworkers and friends.

Histology is not understood by many. It's unfamiliar and unknown. We don't fit in with any other science. We're different. We're unique. The fact is, we are nearly the only science that is completely hands on. Every other field, from the blood bank to chemistry, is almost fully automated and operational. Currently, it is impossible to automate histology. Unlike some other sciences, replacing people with machines is not an option in histology. Histology depends completely on histotechnicians. We have skills. We learn techniques and the proper way of doing things. There is still a great deal of interpretation and discretion amongst histotechnicians when producing their work. No way is wrong, but we have different ways of coming to the same conclusion. We use the best techniques to do the job adequately. Such techniques have been created with great purpose for one reason, to improve the overall quality of histology.

Embedding

2.0 Embedding

Embedding hacks are limited. I've acquired quite a collection over the years. Embedding techniques are clever and require skill. This discipline is very meticulous and requires quickness and organization. Embedding techniques differ from techniques in other histology disciplines. These techniques require more of an innovative attribute, regarding improving the overall process by using these methods. For example, in microtomy, you may have discovered a special unorthodox solution when cutting to improve the quality of sections on slides. On the contrary, when embedding, if the tissue is problematic, you are more limited with the material this discipline offers. New maneuvers must be incorporated because you have only your hands to utilize.

Air Bubbles

2.1   Bubble Trouble

To remove air bubbles that occurred while embedding

Problem: During embedding, air bubbles are inadvertently left in a block, causing holes in the paraffin, negatively affecting the tissue sections.

Solution: Fill the holes with paraffin using a pipette at the embedding center. After solidifying, section the block. You can then file the block or re-embed it for file.

Pro: This is an expeditious way to cut a block, without having to re-embed it.

Con: The block may still require to be re-embed before filing.

Hard Tissue

2.2 Foreign Body/Hard Tissue in a Block

To remove and permanently store uncuttable foreign body or hard tissue in a block

Problem: If a block contains a hard piece of tissue or a foreign body that won't cut at the microtome, here is a good approach.

Solution: You can melt down the block, remove the hard piece of tissue, and place it on the other side of the cassette. Pour paraffin on it and let it solidify for safe keeping if needed for later use. Hard tissue may occur for many reasons. It may be the nature of the tissue. The tissue was not put in formalin quick enough at the grossing station causing crunchy specimens. The tissue is calcified. The tissue sat in the paraffin step of the processor too long, or it is a foreign body.

Pro: The block is much easier to cut.

Con: Make sure you do not remove anything significant that the pathologist may need to see, regardless of how it cuts.

2.3 Polar Ends

Allows hard and soft tissue to be cut simultaneously in one block, without effecting each other's quality

Problem: When cutting hard and soft tissue in the same block, the soft tissue tends to shred, causing horrific sections.

Solution: I recommend embedding the hard tissue on one side of the block and the soft tissue on the opposite side of the block. This way, the soft tissue cuts perfectly and gives a great section, while the hard tissue shreds the block without interfering with the quality of the soft tissue.

Pro: The soft tissue will give a good section.

Con: None

Lost Tissue

2.4   Find & Go Scrape

A method to help find tissue that is lost while embedding

Problem: Tissue is dropped during embedding. I have witnessed this happen multiple times.

Solution: In a circumstance when the tissue cannot be found, the last resort may be scraping the floor. Be careful not to walk too close to the embedding unit or the tissue could stick to the bottom of your shoe. Use a floor scraper for paraffin removal. Scrape the floor in one direction, one long stroke at a time. A group of paraffin and dirt will stick to the scrapper blade. Use a gauze pad to wipe off the dirt. You may need your coworker's help. If you find the tissue, make sure it resembles the missing tissue. Report back to your pathologist, so they can identify if this is the correct tissue.

Pro: This may help you find the tissue, otherwise, impossible to find.

Con: Be careful. You may find other tissue as well.

2.5 Jumper

Prevents the loss of tissue during embedding

Problem: During embedding, if a technician squeezes the forceps too hard, the tissue may shoot and can bounce away. It can travel quite a distance. I refer to this type of tissue as a "jumper." A common "jumper" is skin and cervix. Usually, they can be found with two good eyes. However, sometimes, they are stuck in the paraffin on your gloves. They can be in any nook and cranny on the embedding unit, the floor, or even in your scrub pockets next to yesterday's block!

Solution: When transferring tissue from the cassette to the mold, keep the tissue as close to the cassette as possible and put it directly into the mold. Do not pass go. Do not collect $200 dollars. Once the tissue is in the mold, it can be oriented.

Pro: This helps the tissue from accidentally being shot and lost at the embedding unit.

Con: None

2.6 Meshy

A method for opening a mesh bag with minimal tissue loss and no detached mesh fibers

Mesh bags can be a hassle and a pain in the neck. I never met a histotechnician that likes them. Opening them at the embedding unit is frustrating. Here is an alternative method.

Problem: When opening a mesh bag, it is easy to leave tissue fragments behind. There is also the possibility of accidentally removing mesh fibers.

Solution: You can open mesh bags relatively quickly while scraping all the tissue off without the paraffin solidifying too fast or losing tissue. Open the mesh bag as normal with your fingers and forceps. Then, with one hand, use your pointer and middle finger like scissors to hold the top of the bag. Hold the bottom with your thumb and ring finger. Gently scrap the bag with a small knife towards your body. All the tissue will come off and stick to your knife. You will not get any of the mesh bag threads mixed with your mold. There will be no loose tissue on the embedding unit. Use the forceps to scrape the tissue off the knife and into the cassette mold.

Pro: Less tissue loss and a reduced chance of getting mesh bag fibers in your mold.

Con: It is difficult to maneuver and a little hard to master this technique.

Orientation

2.7 Orientation Station

To help identify the proper orientation of a specimen

Problem: Difficulty orienting a specimen. Trouble identifying the cut side, ink side, epidermis side, etc.

Solution: Embed the specimen to your best ability. Then solidify the block and melt it down. Upon seeing the specimen for the second time, the proper orientation will stand out. Now you can re-embed the block permanently.

Pro: This may help you determine the orientation.

Con: Embedding and re-embedding the same block can be tedious.

Paraffin Prevention

2.8　Baby Oil

Prevents paraffin from sticking to your gloves

One of my pet peeves is paraffin constantly sticking to my gloves during embedding.

Problem: Paraffin's adhesive chemical makeup is interfering with embedding.

Solution: Put a drop of baby oil on your gloves before embedding. The baby oil and the paraffin are not miscible. This will prevent the paraffin from sticking to your gloves.

Pro: There is no more sticky wax everywhere! There is no need to change your gloves frequently.

Con: Do not touch the tissue! You do not want the oil being absorbed by the tissue or you will have bigger problems than waxy gloves!

2.9 Edward Icy Hands

Helps remove paraffin from gloves

I invented this technique. I used to embed and shave hundreds of blocks daily. I needed speed. I started this hack unconsciously, but it works.

Problem: Paraffin buildup all over gloves while embedding.

Solution: Spray your gloves with freeze spray at the tip of your fingers. The freeze spray instantly solidifies the paraffin. Rub your fingers together, and the paraffin crumbles off.

Pro: This keeps you from constantly having to change your gloves. It prevents tissue from getting stuck on your gloves and possibly becoming lost or cross-contaminating specimens.

Con: Order extra freeze spray, so you don't get caught in a sticky situation!

2.10 Square Cold Spot

Prevents paraffin build up on the cold spot

The square cold spot is the small square on the embedding unit where the tissue is oriented. Do not confuse this with the cold plate. I've never heard of it being called a square cold spot, but evidently, that's the name. This is my favorite embedding trick.

Problem: The constant scraping of the cold spot between embedding each cassette.

Solution: When embedding, place a folded Kim wipe entirely covering the cold spot. Get a little flask full of water. Wet the Kim wipe with water using a pipette then embed as normal. Orient the tissue in the mold directly on the Kim wipe. The paraffin won't stick or build up on the cold spot at all! There is no need for scraping anymore. When the Kim wipe starts to dry out, just add a little more water.

Pro: It increases the speed of the work. It keeps the area cleaner, without all the waxy buildup.

Con: The initial set-up is a little time-consuming.

Stability of Blocks

2.11   Refill

To easily add paraffin to an already embedded cassette

Problem: During embedding, not enough paraffin was dispensed into a cassette before it was placed on the cold plate, which may cause instability of the block.

Solution: The embedder just placed a block with insufficient paraffin onto the cold plate. Do not bring the mold back to the paraffin dispenser on the embedding unit. The paraffin will spill. Just take another empty mold and dispense paraffin into it. Then pour the paraffin from the mold into the block on the cold plate that does not have enough paraffin.

Pro: This keeps paraffin from spilling.

Con: None

Tissue Size (Big to Small)

2.12   Magnify

Allows an easy approach to see small tissue specimens at the embedding unit

Problem: Difficulty with tissue orientation at the embedding center. This method is usually used for cervix, skin, and tissue put on edge. For example, you may have trouble identifying the cut side of skin or cervix specimen.

Solution: First, dispense paraffin into the mold. Second, place the tissue inside the mold. The paraffin appears to magnify the tissue. It is easier to identify how the specimen must be oriented.

Pro: Helps easily identify the orientation of the tissue.

Con: None

2.13 Something Blue

Notifies the histotechnician at the microtome that the tissue in the block is tiny

Problem: The histotechnician needs to be made aware of the size of a tiny biopsy before sectioning.

Solution: Here is a way to designate tiny tissue at the embedding station. I've seen technicians cut blue grossing sponges into tiny pieces. They are not always blue. They can be a multitude of colors, but I always think of the blue ones as the primary sponge color. While embedding put a small fragment of sponge on the top of the cassette (the side where the lid goes). Fill the cassette with paraffin. The blue fragment of sponge is visible in the solidified paraffin on the top of the block. This notifies the histotechnician the tissue is tiny when cut at the microtome. Sometimes, grossers write tiny on the side of the cassette. However, some tissue at gross is big, and after it is processed, it shrinks or fragments. The embedder can now note the tissue is tiny for the cutter.

Pro: When the cutter is made aware the tissue is tiny before sectioning, they won't face deep into the block, and there will be no loss of tissue.

Con: This technique is a little time consuming when embedding.

2.14 Tissue House

To embed oversized tissue

This is a great technique. It is so simple, yet unconventional.

Problem: Sometimes, the grosser just cuts the specimen too fat. You can ask them to cut the tissue thinner, and they will for a week but then they go back to chunks. It's ironic when the tissue is cut so thick and smashed so hard that the cassette lid buckles and you can see the imprint of the cassette lid in the tissue. This occurs for many reasons. Sometimes, the grosser just doesn't want to print an extra cassette. In that case, I would cut the tissue at the embedding center and change the gross (especially skin tips that were too tall). However, sometimes, the grosser/pathology assistant may have, for example, a muscle with an attached tumor they don't want cut. They prefer not to separate the muscle from the tumor. They purposely squeeze it into one cassette. Last, the grosser may think the tissue will shrink enough during processing to fit in the cassette, but that is not always the case.

Solution: When embedding, flip the cassette over and place it in the mold upside down. Add paraffin. Then place it on the cold plate. There is more room for the cassette to house the tissue. Try it. It works.

Pro: This hack saves time and a discussion with the pathology assistant.

Con: When the block is filed, the button, which is the part of the tissue enclosed in paraffin, will be on the opposite side of the cassette.

Tissue Stuck to the Sponge

2.15   Taper Saver

To remove tissue stuck to a sponge without getting sponge fragments in your mold

Problem: The tissue at the embedding unit is stuck to the sponge. When you try to pull it off with your forceps, it removes sponge fragments.

Solution: Take your mold and dispense paraffin into it. Place the sponge on the mold with the tissue facing down. Put the mold on the cold square spot. Quickly push down on the sponge with a taper, allowing the paraffin to only partially solidify. Then pull off the sponge. The tissue will now be in the solidified paraffin, not on the sponge. There also will be little to non sponge fragments.

Pro: This is one of the only ways to get stuck tissue off a sponge.

Con: Some of the tissue may not detach from the sponge.

Under Processed Tissue

2.16   Chamber Saver

To complete the processing of under processed tissue

Problem: The tissue in a block is under processed.

Solution: Place the cassette that contains the under processed tissue inside the embedding center wax chamber. Make sure the lid is tightly closed. It will cook in the paraffin and finish processing on its own.

Pro: A rapid way to finish processing a cassette as opposed to reprocessing it.

Con: Do not forget about the cassette!

Uneven Planes

2.17   Re-embedding blocks

Re-embedding faced blocks without additional tissue loss

Problem: A block being faced has tissue on different planes. The technician tries to go deeper to get a full section of all the pieces. However, one piece of tissue in the block is too deep, and the other is too shallow. The block must be re-embedded to obtain a full section of all tissue fragments. Upon re-embedding the block, the tissue thickness of all pieces is inconsistent. The technician wants to make sure when cutting the block, the precut tissue is not cut away.

Solution: When re-embedding the block, first push the uncut tissue as far down in the mold as possible. However, only push the precut thin tissue down a little. When cutting, you will face the thick tissue first and the thin tissue second. You will get a full section without sacrificing tissue.

Pro: No loss of tissue when cutting a precut block.

Con: None

Microtomy

3.0   Microtomy

There are numerous microtomy hacks. Sectioning techniques utilize an array of materials as compared to the other histology disciplines. They use decal, freeze spray, and even soap! Most microtomy hacks are effective. I remember when I first learned to cut. It was like patting your head and rubbing your tummy simultaneously. Your body is just not coordinated for it, but over time, it's like riding a bike. However, occasionally, you might hit a bump in the road. Here are hacks to keep you going smoothly.

Bubbles in a Paraffin Ribbon

3.1   Bubble Bath

To prevent air bubbles in paraffin ribbon in the water bath

Problem: Air bubbles in the water bath are interfering with tissue sections.

Solution: Use deionized water. DI water is water purified by distillation. The theory is it helps minimize air bubbles.

Pro: Reduces air bubbles.

Con: None

3.2 Bubble Pop

Popping air bubbles in a paraffin ribbon

Problem: Air bubbles are in the tissue of a paraffin ribbon floating on the water bath.

Solution: Carefully take a curved pair of pointed forceps and pop the bubbles.

Pro: No need to cut another section.

Con: Make sure the forceps are free of debris or your ribbon will be ruined.

3.3 Penny Drop

To prevent air bubbles in a water bath

Problem: Too many air bubbles in the water bath are interfering with the tissue in the paraffin ribbon.

Solution: Place a penny in the water bath. The copper helps reduce air bubbles. Pennies keep fountains and bird baths clean due to copper's anti-bacterial effects. If you don't have access to deionized water, use the penny in tap water instead.

Pro: Decreases amount of bubbles in the paraffin ribbon.

Con: The theory is still undetermined.

Cysts

3.4 Cyst Concealer

Conceals smelly cyst odor

This is a great hack for those smelly cyst blocks.

Problem: After cutting a cyst block, the block smells.

Solution: After cutting, dip the button of the block in the paraffin to conceal the odor. The button is the part of the block that contains the tissue. The last step is to solidify and file.

Pro: The cyst odor is concealed.

Con: None

3.5 Vicks Vapor Rub

To cover up smelly cyst odor at the microtome

Problem: Gross, smelling cyst odor.

Solution: Put a tiny amount of Vicks Vapor Rub under your nose. This is a common practice in histology when doing autopsies.

Pro: Covers up the cyst smell.

Con: Some dislike the smell of Vicks.

Decal

3.6 Backflip Blocks

Prevents blocks in decal from flipping upside down

Problem: When putting multiple blocks in decal solution at once, they constantly flip over and float to the top. The tissue is no longer emerged in the decal.

Solution: Use a shorter and wider container and fill it approximately 1 cm high with decal. The blocks will float on top and will not flip over.

Pro: The tissue will be entirely submerged in decal solution.

Con: None

3.7 Lazy-Cal

Saves time when removing a block from decal

After soaking blocks in decal or any other solution, the block must be rinsed off in water before to cutting.

Problem: It's not so much a problem, as much as an inconvenience, to get up to rinse your block off in the sink after taking it out of decal solution.

Solution: After using decal solution, remove the block. Use gauze to remove the excess decal and dip the block in your water bath to avoid a trip to the sink. The decal is too minuscule to affect the water in the water bath.

Pro: Avoid a trip to the sink.

Con: None

3.8 Save the Decal

To not waste decal

Problem: Only a tiny amount of decal is needed to soak a block.

Solution: Flip your block over on the counter and pipette the decal solution on top of the paraffin button. It won't leak off, and it will cover the whole surface area until you need to cut it. It's kind of like, in grade school, when you pipette water droplets on a penny to see how many it can hold. It's an easy hack if you don't have a container of decal solution and you just need to soften one block.

Pro: Saves decal! The decal can be pipetted back into the bottle for reuse.

Con: Only works with one or two blocks. If you do this with multiple blocks, you will waste decal.

Dry Blocks

3.9   Moist Towelette

For dry blocks lacking moisture

Problem: Horrible dry sections with holes when ribboning.

Solution: Wet a paper towel with the warm water from your water bath and place the faced block on the wet paper towel. After the block soaks for a few minutes, rest it on the cold block then cut. This is an alternative to floating your block on your water bath for added moisture (3.11 Warm Dip).

Pro: Does not leave debris in the water bath.

Con: None

3.10 Moisturize

Adds moisture to dry blocks

Problem: The blocks are dry and not sectioning well.

Solution: Add warm water to the blocks for moisture. When cutting, dip your finger or gauze into your water bath and rub it on the block in your chuck. It will give the block moisture. The block will be less dry, and your ribbon will stay intact.

Pro: Works rapidly when moisture is needed.

Con: The issue may not necessarily be a moisture issue.

3.11 Warm Dip

Adds moisture to bloody blocks

Typically for bloody specimens. I learned this at Colombia University's histology laboratory in New York.

Problem: During sectioning, a block is dry, falling apart, and getting holes in the paraffin ribbon while facing.

Solution: After facing, float the block on your warm water bath for approximately 30 seconds and put it on ice before cutting. This will provide extra moisture, so the block is not too dry when cutting. This is the hack I use the majority of the time. It works considerably well and will give you a perfect section.

Pro: Produces satisfactory sections and is simple to perform.

Con: It may make your water bath a little dirty. Make sure you clean off your block before floating it on the water bath.

Hard Tissue

3.12   Liquid Fabric Softener

To prevent static and soften the tissue before cutting
This hack has more than one purpose.

Problem: There is a static charge, lack of moisture, and hard tissue interfering with block cutting at the microtome.

Solution: Dilute the laundry liquid fabric softener in water and place the faced block in it for about 10 minutes. The fabric softener will reduce static, add moisture to the block, and soften the tissue. The friction between the tissue and the blade creates static electricity neutralized by the fabric softener.

Pro: Softens the block.

Con: None

3.13   Lotion

To soften nails and hard specimens

Problem: Nails are difficult to cut. This is an alternative to Nair (3.17 Nair It).

Solution: Squirt lotion on your faced block. Leave it on as needed.

Pro: Helps soften nails and hard tissue.

Con: You may still have to use additional softner depending on the hardness of the tissue.

3.14 Pop Goes the Block

When cutting a piece of hard tissue, it pops out of the block

Problem: When a piece of tissue pops out of a block, it usually happens repeatedly. Normally, you re-embed it. Next, you face it. Then you soak it in the surface decal. However, if it still pops out, here is a fix.

Solution: Don't re-embed it immediately. Put the piece of tissue that popped out directly in decal for 30 minutes. If needed, melt the rest of the tissue and place it in the decal solution too. Then, rinse it. Wipe it dry with a paper towel. Re-embed it and cut. You probably will not have to use surface decal when facing.

Pro: It is actually, much faster than continuing to cut a block with the tissue constantly popping out.

Con: None

Identification of Block History

3.15   Sharpie Solution

Identifies and documents the histotechnician that cut previous blocks

Problem: Identifying the histotechnician that sectioned the blocks.

Solution: I was taught this hack at a private dermatology laboratory. It identifies what specific blocks each technician cut. Each technician is designated a different Sharpie color. After the blocks are cut, draw a line with your Sharpie down the right side of the blocks in the box holder. Later, if there is an issue with a block, for example, the tissue is put on the wrong slide, you can easily identify what technician cut it. Also, if you must recut a block, you will know which microtome it was cut on. Thereby, you can cut the block at the same angle, assuming the technicians use designated microtomes. It sure beats writing down all the accession numbers!

Pro: Allows the technician to be identified. This also allows the block to be recut at the same microtome. There are a multitude of Sharpie colors.

Con: In a laboratory with multiple technicians, you may have to use an abundance of different colors of Sharpies. A color key chart can be made to minimize any confusion.

Microtomy Station

3.16   Lockless Microtome

No blade guard on the microtome

Problem: Some of the old microtomes do not have blade guards. If you must keep a partially used blade, here is a quick fix.

Solution: Using a Kim wipe, place it vertically. Put a hole in one end by pushing it through the tissue block clamp, the metal piece that sticks out of the top directly above the chuck holder. Leave it there. This notifies a histotechnician there is a blade in the holder.

Pro: Very easy and not time-consuming.

Con: None

Nails

3.17 Nair It

Soften nails prior, to sectioning

Problem: Nails are often hard and need to be softened before and after processing to be sectioned properly.

Solution: At gross, place the nail on a small grossing sponge in a cassette. Apply Nair, covering the nail. Let it sit until it softens. Scrape off the Nair with a scalpel. Put the nail between the two grossing sponges in the cassette and leave it in formalin to be processed. Make sure not to lose any tiny nail fragments in the Nair. At the microtome, face the block then reapply Nair on the block for approximately 20 minutes. Wash the block off in water, then cool and cut. The nail is easier to cut, less likely to pop out, sections better, and will stay on the adhesive slide. The Nair breaks down the keratin. Keratin acts as a protective shell for the nail. The Nair dissolves the disulphide bonds between the keratin proteins in the nail. It acts the same way it does when using Nair on hair, penetrating and softening the nail.

Pro: This method produces a good nail section.

Con: The Nair procedure may prolong the turn-around time of the specimen for approximately twenty-four hours. Nair has a pungent odor.

Optimal Water Bath Temperature

3.18 On the Rocks

To reduce the water bath temperature quickly

Problem: The water bath is accidentally heated too hot.

Solution: If your laboratory works with ice, just drop two cubes in the water bath. This will instantly cool it down.

Pro: Works instantly.

Con: None

3.19 Whistle While You Work

Allows optimum temperature of water bath immediately

This hack is a genius, yet so simple. It's very frustrating to wait for your water bath to heat prior, to cutting. It's also very annoying when your coworker arrives first and decides not to fill your water bath. This strategy keeps you from having to heat your water bath in the break room microwave!

Problem: Waiting for your water bath to heat before cutting.

Solution: Use an electric tea kettle. Fill your water bath with approximately 1/3 hot water from the kettle and the rest with room temperature deionized water. Your water bath is now ready for use.

Pro: Heats your water bath for immediate use.

Con: Make sure you don't put too much hot water, or you will be waiting for it to cool down.

Ribboning Issues

3.20   Finger Press

A paraffin ribbon will not form when cutting

Problem: When cutting, each section keeps breaking off, not allowing a paraffin ribbon to form.

Solution: Lock your microtome in place and carefully rub your finger across the bottom of the button of the block in the chuck. Your finger will provide just enough warmth to allow the paraffin **to stick together to** achieve a ribbon.

Pro: Works great.

Con: None

3.21 Pencil Unsharpener

Dulls a new microtome blade

Problem: The brand-new microtome blade is so sharp you can't obtain a paraffin ribbon.

Solution: Dull a microtome blade. If your blade is too sharp and you can't get a ribbon, then secure the blade in the microtome and rub a pencil eraser over the blade. I discovered rubbing gauze over the blade works too. I prefer gauze because you will not get any knife marks. This works with all gauze, whether it is soft or rough. Sometimes, I still use this technique if I'm in a jam. This works well on the tiny tissue when you can't afford the loss.

Pro: Helps the histotechnician obtain a paraffin ribbon.

Con: This technique may dull the blade a little too much, causing you to have to replace it.

Soggy Tissue

3.22 Cling Wrap

Prevents blocks on the cold tray from becoming overly saturated with water

Problem: Blocks become over saturated with water on the ice tray prior to cutting, making the tissue a little mushy.

Solution: Put Cling Wrap on your ice tray and your blocks on top. The Cling Wrap acts as a barrier to prevent water absorption. The blocks will still get cold before you section them, without becoming overly wet.

Pro: Keeps blocks cold and not too moist. This will prevent multitaskers from having to flip their blocks over, keeping them from siting on the watery ice tray too long and becoming too saturated.

Con: Cling Wrap can be clingy and hard to maneuver.

3.23 Foil

Prevents blocks on the ice tray from becoming overly saturated with water

Problem: Blocks become over saturated with water on the ice tray before cutting. For those of you who do not like your blocks overly wet, this hack is for you.

Solution: Put aluminum foil on your ice tray and place your blocks on top. The aluminum acts as a barrier to prevent water absorption. The blocks will still get cold before you section them without becoming overly wet.

Pro: Keeps blocks cold and not too wet.

Con: The blocks may be too dry for some histotechnicians.

Static

3.24   Aluminum

To prevent the dreaded static cling (Aluminum)

Problem: Static can be frustrating when cutting. You may cut a perfect ribbon, but the ribbon will stick to your forceps and get ruined due to static. It doesn't matter if you wipe the chuck holder with gauze, the static will not go away.

Solution: Aluminum foil decreases static. Place it behind the blade holder. How it works: During cutting, the rubbing of the block with the blade builds static. There is an exchange of electrons. The electrons become positively and negatively charged, creating static. The foil discharges any static buildup and keeps the electrons separate, which decreases static and will speed up the cutting process.

Pro: Reduces static.

Con: None

3.25 Damp Cloths

Reduces static when cutting at the microtome

Problem: Static is interfering with cutting at the microtome.

Solution: Place wet Kim wipes in your catch tray. They can be placed over the paraffin shavings. The damp Kim wipes will reduce dryness and provide moisture reducing static. The damp Kim wipes make the air humid. The air becomes less dry. The electrons will reunite with the molecule missing an electron, and the static will be reduced.

Pro: Easy to do.

Con: The Kim wipes will make the catch tray slightly wet. The tissue and paraffin may partially stick to it when you wipe it out.

3.26 Dryer Sheets

To stop static before it starts

Problem*:* Static is interfering with sectioning blocks at the microtome.

Solution: Place a dryer sheet behind the blade. Fabric softener sheets are positively charged material that balances out the charges, reducing static charge.

Pro: Reduces static charge quickly.

Con: Over time, you must replace the dryer sheet with a new one.

3.27 Humidify

Static again!

Problem: If you have an overabundance of static daily, this may be a remedy for you.

Solution: As your last resort, buy a small humidifier. The moisture it produces helps reduce static charge by making the air less dry.

Pro: Helps reduce the static.

Con: It may make noise or take up space.

Tissue Placement on Slide

3.28 Glue Trick

To create adhesive slides

Problem: You run out of positively charged slides.

Solution: Add a drop or two of glue (Elmer's glue) to your water bath. The glue will allow the paraffin and tissue to adhere to the slides. The ribbon can now be placed in any position on the slide. The ribbon will not move around. This is especially good for Immunohistochemistry controls because one section of the tissue must be mounted to the slide in a particular position.

Pro: Works well.

Con: None

3.29 Quicker Picker Upper

Effortless methods to pick up paraffin ribbons on slides

Problem: Sometimes, picking up sections with slides on the water bath can be difficult. Often the tissue runs from the slide like it's a rat race. This is because the slide, paraffin, and tissue are negatively charged. They repel each other.

Solution: There are multiple ways to pick up tissue and center it on the slide. First, lay down your ribbon and break off a section. Use a pair of forceps to drag the corner of the section on to the slide. It will adhere nicely. A second method is to touch the tissue to the top of the slide. If there is a brand name or logo at the top of the slide, the paraffin will stick to it. A third method is to touch the paraffin slightly onto the side of the slide. The section will attach. Fourth, break up your sections with your forceps or fingers. Then pick up the tissue with the slide upside down.

Pro: Helpful ways to pick up a paraffin ribbon.

Con: None

3.30 Sliding on a Slide

To adhere stubborn tissue to a slide

Problem: When you get a perfect section on your slide, but the section keeps sliding around. You don't want it to fall off or slide too far to one side or to the bottom.

Solution: If you blow on the section, it will adhere to the slide. Sometimes, a gentle blow works better, and other times, a forceful blow is more effective. Sometimes, when lifting your slide out of the water bath, your tissue slides down. You can blow on the section immediately as you slowly lift the slide out of the water, and it will adhere to the center of the slide.

Pro: Works well to center the tissue on the slide.

Con: None

3.31 Slip & Slide

To separate two blank slides stuck together

Problem: Slides are stuck together due to moisture. This is frustrating. They are very hard to pull apart.

Solution: Drop the attached slides into your water bath. You may continue cutting your blocks. In approximately 2 minutes, take out the slides. They will separate themselves.

Pro: The slides will detach.

Con: None

3.32 Two Sections on One Slide

Placing two sections on one slide

Problem: If you must put two sections on one slide, which I hate, here are two ways.

Solution:

Vertical: Leave your ribbon attached to the water bath and pick up the ribbon vertically on the left side of the slide. Then pick up the second ribbon on the right of the slide at a slight angle. The ribbon will stay on better if the two ribbons touch each other in the middle of the slide.

Horizontal: This is the easiest approach if the pathologist wants 3 levels on the slide. Lay three ribbons down. Pick up one or two small sections at the top of the slide so it runs horizontally across the slide. Do the same for the middle and end of the slide.

Pro: Efficient ways to pick up two levels on one slide.

Con: None

3.33 Upside Down Slide

An easy way to pick up control tissue

Problem: Picking up control tissue with an adhesive slide can be frustrating. If it touches the wrong part of the slide, the section is ruined. The tissue must be mounted to the appropriate position.

Solution: Picking up the section with the control slide upside down seems to ease the process. The slide does not have to go as deep in the water. This leaves less surface area for error.

Pro: Adheres the tissue section to the exact position.

Con: None

Under Processed Tissue

3.34   Frozen-less Section

Gives an almost impossible section on an under processed block

Problem: This hack is exclusive for under processed tissue that will not ribbon. I learned this at a private dermatology laboratory in Florida. This is for a section of tissue that partially falls apart to the extent that the ribbon cannot be transferred from the chuck to the water bath.

Solution: Treat the procedure almost as if you would a frozen section. Cut a ribbon and hold it with your forceps keeping the other end attached to the chuck. Do not move it. Wet your slide and bring it to the chuck. Carefully adhere your section to the slide as you would a frozen section. However, put the slide under the ribbon and carefully lift the section up. The section is usually wrinkled so you must quickly dip the slide into the water bath. Be careful the tissue does not wash off the slide.

Pro: Gets an otherwise impossible section.

Con: You may want to reprocess the block unless otherwise instructed.

3.35 No Explode Tissue

Allows an acceptable section on an under processed block

Problem: The tissue is under processed.

Solution: Freeze spray your water bath right before you lay down a section. This will reduce the temperature of your water bath for an instant and delay the section from disintegrating. Sections of tissue disperse when placed in warm water because, when you heat water, it expands. The section diffuses. In addition, the tissue dissolves better in hot water because the molecules are not as compact. They travel at a rapid rate and utilize more energy.

Pro: Keeps tissue from exploding when placed in the water bath.

Con: A great number of technicians frown upon the use of freeze spray due to tissue artifacts. Artifacts may resemble cancer cells microscopically.

3.36 Suds in a Water Bath

Allows an adequate section to be cut on an under processed block

This hack is phenomenal. It constantly works on the most under processed blocks. I learned this from a knowledgeable traveling histotechnician.

Problem: The tissue is under processed and deteriorates when placed in a water bath.

Solution: Put a drop or two of liquid soap on your fingers and rub them together in the water bath. Lay your ribbon down on the water bath. The under processed tissue will stay intact. The soap decreases the surface tension of the water and increases the viscosity of water. The higher the viscosity, the more your section will float momentarily, making this a successful hack.

Pro: Works phenomenally with under processed tissue to give you the best section possible.

Con: The soap only lasts temporarily so you may have to add more.

Wrinkles

    1. Prevention

3.37   **Alcohol Bath**

    Prevents wrinkled tissue sections

    Problem: Wrinkles are in paraffin ribbon floating on the water bath.

    Solution: Add a capful of Alcohol over your cold tray or to your water bath. This will help get rid of wrinkles before they start. Alcohol and water are miscible. The Alcohol falls between the water molecules and pushes the oxygen away, keeping out the wrinkles.

    Pro: Helps keep wrinkles out of the tissue.

    Con: None

3.38 Ammonia

Prevents wrinkled tissue sections

Problem: Wrinkled tissue sections in the water bath.

Solution: Add a capful of Ammonia over your cold tray. This will help get rid of wrinkles before they start. Ammonia forms hydrogen bonds and is miscible with water.

Pro: Helps keep wrinkles out of the tissue.

Con: It is has a strong odor.

2. Reduction

3.39 **Double Dipping**

To remove wrinkles from a paraffin section

Problem: The paraffin section floating in the water bath has wrinkles.

Solution: Obtain a container of room temperature water. It does not have to be big, just big enough to lay down a ribbon. First, lay your ribbon on the water bath. If it is wrinkled, then pick up a section or multiple sections with a slide. Place the slide in the container with the room temperature water. Let the ribbon float freely off the slide for approximately 30 seconds. Then pick up the section with the slide and let the section float in the warm water bath. Watch the wrinkles magically disappear! The hot water expands the tissue and the cold water constricts the tissue causing the wrinkles to be stretched out.

Pro: There is no need to resection the block.

Con: One must be careful not to ruin the paraffin ribbon when transferring it from one bath to another.

3.40 Runway

Removes wrinkles in a paraffin ribbon

Problem: Wrinkles are in a paraffin ribbon floating on the water bath.

Solution: Take your ribbon from your microtome. With your forceps, drag it through the air with it coming towards yourself until it touches down on your water bath. The paraffin will stretch out in the air, and this helps smooth out the wrinkles.

Pro: This works well, especially on fatty tissue or tissue in big molds.

Con: None

3.41 Wrinkle Release

Removes wrinkled tissue from a paraffin ribbon

Problem: Wrinkles are in paraffin sections floating on the water bath.

Solution: Use forceps to remove wrinkles in a paraffin ribbon. For small wrinkled tissue floating on the water bath, gently tap around the wrinkles with the tips of your curved forceps. This will take out the wrinkles. Make sure the tips of your forceps are free from debris.

Pro: Removes wrinkles. Additional sectioning of the tissue is unnecessary.

Con: If you have debris on your forceps, your paraffin ribbon will be ruined.

Frozen Sections

4.0 Frozen Sectioning

Frozen Sections are a very important part of histology. They can be frantic. They require organization, speed, and skill. The surgeon is dependent on the pathologist, and the pathologist is dependent on the histotechnician for good sections and quick turnaround time. The longer the patient is under anesthesia, the more troublesome. The pathologist must give the surgeon quick results. Here are techniques to help aid you.

OCT Compound

4.1   Frozalin (Frozen to Formalin)

To safely rinse off OCT

The worst thing a histotechnician can do during a frozen section is to lose the tissue down the drain. Many pathologists prefer the permanent H&E to the frozen section H&E because it gives a better section and a more secure diagnosis.

Problem: Rinsing OTC (optimal cutting temperature compound) off the fresh tissue directly after a frozen section can be unsafe.

Solution: Don't do it! There is no need. Just put the tissue with the OCT on it directly in formalin. The OCT will disintegrate in the formalin.

Pro: The tissue will not go down the drain.
Con: None

Sectioning

4.2   Fog It Up

Keeps tissue from jumping on your slide during a frozen section

Problem: A common problem when cutting a frozen section is that the tissue adheres too quickly and improperly on the slide. Sometimes, when you pick up the tissue section with the slide, the tissue "jumps" onto the slide, ruining the now wrinkled and folded section.

Solution: Blow on the back of the slide in a huff to fog it up. The humidity will prevent the tissue from adhering too quickly.

Pro: A very simple solution.

Con: None

Static

4.3　Cold Static

Prevents static in cryostat

Problem: Static is in the cryostat, which is preventing a section of tissue from being picked up by a slide.

Solution: Wet a paper towel and lay your blank slide on it. Then wipe the back of the slide with the wet paper towel.

Pro: Helps prevent static.

Con: None

4.4 Cry-O-Static

Prevents static in cryostat

Problem: Static is occurring and interfering with obtaining a tissue section when cutting on a cryostat.

Solution: Place a dryer sheet in the cryostat under the blade (works with the microtome too).

Pro: Prevents and reduces static.

Con: None

Grossing

5.0 Grossing

Grossing hacks are predominantly aimed at keeping the overall final sample of tissue at its ideal best. The slides produced will be of the highest quality. Grossing hacks aid the histotechnician at the embedding unit, which helps the cutter at the microtome. It trickles down. It has a domino effect.

Grossing is the first step in the histology cycle. At gross, it is important for the pathology assistant/grossing technician to communicate with the histotechnician. The communication should consist of the tissue samples received and the orientation. When applying these grossing techniques, overall, there will be less loss of tissue and proper consistent orientation. These are a few simple techniques a pathology assistant or grossing technician can do to help improve the overall quality of the final result of the specimen.

Cleaning & Disinfecting Grossing Utensils

5.1   Dreft

An easy way to clean grossing instruments

Nothing is worse than grossing with unsanitary, unclean scalpels, forceps, and other grossing utensils.

Problem: Grossing utensils, such as forceps, probes, and blade holders, get very dirty. Over time, the gunk builds up.

Solution: I worked with a pathologist who learned in medical school to soak the grossing utensils overnight in baby laundry soap diluted with water. The next morning, the instruments are clean, with no residue.

Pro: No scrubbing of utensils involved.

Con: The detergent can be rather expensive and cleans but does not sanitize.

Epidermis Error

5.2   Ellipse Skin Fragment

Allows the grosser to approximate a measurement of an ellipse skin specimen after it is dissected

Problem: When I worked at a private dermatology laboratory, I grossed hundreds of ellipse skin fragments daily. Seldom after you gross a specimen, you realize you accidentally forgot to measure it. However, it happens.

Solution: If it's already dissected, you can measure the ink stain left on the paper towel. It should be in the shape and size of an ellipse. If there is no stain, you can always piece the tissue back together if you can recognize the proper placement of the skin.

Pro: You do not have to piece the whole specimen back together to measure it.

Con: Unfortunately, the measurement will be an approximation.

Filter

5.3 Cap Full

An easy way to filter small specimens without using a filter or filter paper

My former supervisor taught me I could filter small specimens (20 ml specimen bottle) using only the cassette, grossing sponge, and the specimen bottle lid.

Problem: Not filtering specimens efficiently.

Solution: First, put the sponge in the cassette. Next, put the cassette on the open side of the lid. The specimen bottle is 20 ml, and the formalin in it is 10 ml. The lid holds approximately 10 ml. Slowly pour the formalin into the cassette, and the sponge will act as a filter. When done filtering, pour the remaining formalin in the lid back in the specimen bottle.

Pro: Works great without a funnel or filter paper.

Con: Pour slowly, or the sponge and cassette will overflow with formalin.

5.4 Forma-Dip

An effortless way to open mesh bags at gross

Problem: Mesh bags are difficult to open.

Solution: They open easier when wet. Dip them directly into the formalin in the specimen container at gross, and they will be more manageable.

Pro: The mesh bag is easier to open. Dipping them in the specimen bottle is convenient.

Con: None

5.5 Squeeze It

Removes all tissue fragments from specimen bottle when tissue is submitted entirely at gross

This is a very effective hack. I learned this from the Medical Director at my former hospital.

Problem: Making sure fragments are not left behind when entirely submitting a specimen.

Solution: Use a squeeze bottle containing formalin. When filtering the specimen, squirt the formalin into the specimen container and the formalin will remove every speck of tissue.

Pro: No tissue fragments will be left in the specimen bottle.

Con: None

Prevention of Cross Contamination, Inaccuracy, and Disorder of Specimens

5.6   Baskets

To keep pre-grossed specimen bottles and cassettes organized to prevent error or loss

When accessioning, many laboratories line up their specimen bottles with their corresponding cassettes. This method is efficient for several laboratories, specifically those that gross an abundance of small specimens. It is beneficial because this promotes speed and allows for the specimens to be checked and confirmed for accuracy typically by more than one person. However, in a hospital setting, there are multiple types of specimens ranging in size and order. Some specimens have few cassettes, and others have a surplus. In this case, the basket method is ideal.

Problem: Loss of specimen bottles and incorrect cassettes.

Solution: Use plastic baskets or containers to hold each specimen with their cassettes. This helps keep the specimens and their cassettes together and keeps the specimen bottles from falling on the floor.

Pro: Each specimen is safely isolated with their identifying cassettes and is easily accessible and mobile.

Con: The cassettes will not be in numerical or alphabetical order.

5.7   Colors of the Eight Pack

Designate specific colors for specific specimens

Problem*:* Accidentally mixing up same site specimens as gross.

Solution: For example, on average, you receive 5 breast cores a day from different patients. Ink each patient's specimen with a different color. Breast Cores – patient one green, patient two blue, patient three red, patient four orange, patient five yellow. This will keep accuracy and verification. If there are multiple grossers, communicate the color ink used, so it is not duplicated.

Pro: Helps maintain accuracy amongst specimens.

Con: None

5.8 Even & Odds

Prevents cross contamination of specimens at gross and during embedding

Problem: Cross contamination of specimens during grossing and embedding.

Solution: When identical specimens must be accessioned back to back, ink every other specimen.

Pro: This will prevent cross contamination of identical specimens at gross and floaters at embedding.

Con: None

5.9   Histo-Pic

For documenting the number of blocks before grossing and or embedding

Problem: To avoid misplacement of cassettes.

Solution: Take a picture of the cassettes in their basket before embedding and processing. This isn't much of a hack, but I have seen it save people from being blamed for losing a block they never received. I learned this at Duke University's histology laboratory in North Carolina.

Pro: Protects the histotechnician that may be blamed for losing a missing block. Helps locate the last place where the missing block may have been.

Con: Just remember you cannot take a picture of the blocks if they have a patient's name on them or anything that violates HIPPA unless it is a work approved camera used for this purpose.

5.10 Triple Check

Confirms the patient's name and site when a specimen must be reviewed for further diagnosis

Problem: Confirming a specimen was labeled correctly.

Solution: At gross, ink the bottle, requisite (the paper), and specimen the same color. For example, if grossing an Antrum specimen, ink the actual Antrum specimen, the requisite, and the specimen bottle with green ink. If there is a discrepancy with a specimen, pull the requisite, specimen bottle, and the block to verify they are all inked green.

Pro: It provides confirmation that the biopsy, requisite, and specimen all match the bottle.

Con: None

Safety

5.11   Safety Blade

Allows a scalpel blade to be removed safely

Problem: Removing a scalpel blade safely. This is a well-known hack. I know a pathologist who cut her finger removing a scalpel blade from a blade holder. Take the proper precautions. A metal blade remover can be helpful; however, most labs do not have them.

Solution: One suggestion is to use your forceps to remove the blade without it ever touching your gloves. Gently release the bottom portion of the blade with the forceps and carefully slide the blade off. Then dispose of it in a red biohazard bin while holding it with the forceps.

Pro: This is a safer approach when removing a blade than using your hands.

Con: None

Tiny Specimens

5.12   Dip Stick

Dye tiny specimens to make them more visible

Problem: Tiny specimens put in mesh bags can be hard to see with the naked eye.

Solution: After placing the tiny fragment(s) in a mesh bag, dip the bag in a container of Eosin. This will stain the fragment(s), so when it comes out of the processor, the tissue will be faintly orange.

Pro: Quick and simple approach for more visible specimens.

Con: The Eosin may be a little messy if you are not careful.

5.13 Orange is the New Black

To make tiny specimens more visible at the embedding center

Problem: Tiny specimens, such as those measuring approximately 0.1 cm to 0.4 cm, are difficult to see, especially when embedding.

Solution: Pipette a drop or two of Eosin on the tiny tissue fragments while grossing. This will stain the tissue light orange, making it more visible. Typically, it may be ideal to put the tissue on a blue grossing sponge before applying. Grossers may use ink instead, but when the specimen is very tiny, it may cling to the wooden applicator, resulting in loss of tissue.

Pro: The tissue will be more visible at the embedding center. There is no need to discolor the tissue processor by adding Eosin to the Alcohol step.

Con: None

5.14 Orange Options

To make small tissue visible with options other than Eosin

A. Alcian Blue-A very common, reliable blue dye that stains polysaccharides. Used at gross. Pipette a drop on each fragment of tissue.

B. Hemotoxylin-Purple in color and stains nuclei. Used at gross. Pipette a drop on each fragment of tissue.

C. Mercurochrome-A over the counter topical antiseptic used for cuts and scrapes. It is an orange/red dye to be used at gross. Pipette a drop on each fragment of tissue.

D. Methylene Blue-A blue dye and medication with many uses. It is added to Alcohol during processing.

E. Mrs. Stewart's Bluing-A blue dye used to whiten fabrics since the 1800s. Used at gross. Pipette a drop on each fragment of tissue.

F. Phloxine B-A red dye used for coloring drugs and cosmetics used at gross. Pipette a drop on each fragment of tissue.

G. Safranin-Water soluble, stains red, and is used in Gram stains. Use in 95% Alcohol on the tissue processor.

H. Toluidine Blue-Stains multiple structures blue and purple. Used in the cassette rack at grossing.

Conclusion

6.0 Conclusion

These hacks are developed to provide superior quality in histology. These techniques have never been officially tested. They have been executed innumerable times in various laboratories worldwide and have yet to be problematic. Pathologists seem satisfied with the outcome these hacks provide. But proper precautions must be put in place and adhered to regarding histology hacks.

A colleague of mine tried to expedite the process of cell blocks. After combining the precipitant with the heated histology gel, the solution regularly goes in the refrigerator to solidify. Experimentally, she put it in the freezer to speed up the process. The cold temperature negatively affected the cells. Under the microscope, the cancer cells were visible but not nearly as pronounced. When re-administrating the initial procedure, the slides microscopically showed very prominent cancer cells. This an ideal example of why these hacks should be tested, before use.

Some procedures that require specific temperature ranges should never be changed. Special stains are an example of a discipline that should not be altered. I worked in a gastrology laboratory, where we used Schiff's reagent for our special stains. After opening the Schiff's reagent, we stored it in the refrigerator. This is common practice in histology. When Joint

Commission inspected our laboratory, a complication arose regarding the storage of the Schiff's reagent. They were puzzled why we stored our Schiff's reagent in the refrigerator. We explained the life expectancy of Schiff's last longer in the refrigerator. However, this manufacturer of the Schiff's reagent labeled their bottles to be stored at room temperature. Joint Commission was concerned how the temperature of the Schiff's reagent would affect the patient tissue. However, after some deliberation, Joint Commission agreed with the proper documentation from the manufacturer that we could store our Schiff's reagent in the refrigerator.

It's immaterial whether or not a practice has been going on incessantly. It is necessary to assess these histology techniques prior to use. However, it is not required to validate a technique. With persistence and repetition, these methods are successful. They give insight on how histology products can be improved. They have allowed us to produce better sections for diagnosis. Histotechnicians are doing independent research subconsciously. These grossing hacks provide the grosser with more insight in histology and allow better communication between the grosser and the histotechnicians. This provides the histotechnician better correspondence with the pathologist and allows for a better understanding of histology. From grossing to embedding to microtomy to diagnosis, an overall improvement in histology is evident.

www.ingramcontent.com/pod-product-compliance
Lightning Source LLC
Chambersburg PA
CBHW07030123O526
45470CB00002B/666